# A SPACE-AGE PICTURE OF THE UNIVERSE

## IN

## LETTERS TO MAUREEN

# A SPACE-AGE PICTURE OF THE UNIVERSE

# IN

# LETTERS TO MAUREEN

With astronauts taking snapshots of one another on the Moon and an orbiting telescope taking detailed photographs of the most distant galaxies, isn't it about time we took a look at a fresh new picture of our universe?

BY

James T. Kane

Copyright (C) 1995 by James T. Kane

All rights reserved under U.S. copyright laws International and Pan-American Copyright Conventions. Alternate name: uni-vers: rotating unit.

ISBN: 1-58500-278-X

# About the Book

This unassuming volume, written by a father, gives answers in plain English to his young daughter's questions about cosmology. It addresses central unanswered questions in the field - the nature of the universe and the cause of gravity -- and offers answers that have not yet been discovered by the establishment of academic astronomers.

All kinds of people will find this book entertaining and interesting: from young students who want to learn basic principles that govern our solar system, to professional scientists who want to explore new theories and ideas about the universe. The book starts with basic principles of physics and space and shows how, through creative thinking, these can be extended to explain some of the central mysteries of our universe.

# INDEX

| | |
|---|---|
| Dedication | iv |
| Introduction: Letter to Maureen | 1 |
| Engineers' Grace | 2 |
| Our Earth | 3 |
| The Moon | 5 |
| Origin of the Moon | 6 |
| The Planets | 8 |
| Spacing of the Planets | 9 |
| Table 1 | 10 |
| Solar System | 13 |
| Centrifugal Force | 16 |
| Attraction | 17 |
| More on Attraction | 18 |
| Attraction Exists Beyond the Solar System | 19 |
| Rotational Energy | 20 |
| Conservation of Rotational Energy | 21 |
| More on Conservation of Rotational Energy | 22 |
| Rotation and Attraction | 23 |
| Elliptical Galaxies, Evidence of Rotating Universe | 24 |
| Rotating Universe | 25 |
| The Path of a Comet | 26 |
| Where Does a Comet Get Its Spin? | 27 |
| Galaxies | 28 |
| A Galaxy of Galaxies | 29 |
| Our Rotating Galaxy | 31 |
| What Keeps Galaxies Apart? | 34 |
| uni-verse: rotating unit | 35 |
| What are Quasars? | 36 |
| Energy is Attracted by Gravity | 37 |
| Energy Waves Attracted by Energy Waves | 40 |
| Matter to Energy::Energy to Matter | 41 |
| Energy Becomes Matter | 42 |
| How is Energy Converted to Matter? | 43 |
| Mattergy | 45 |
| Expanding Universe | 46 |

| | |
|---|---|
| Regenerating Universe | 48 |
| Agglomeration | 50 |
| Spin | 51 |
| Gravity ! | 53 |
| Beyond ? | 54 |
| O, To Create! | 55 |
| How is this Universe Powered? | 56 |
| Creation | 58 |
| Conclusion | 63 |
| Future | 66 |
| Dear Maureen | 67 |
| Acknowledgments | 68 |
| Appendix 1:   Red Shift--A New View | 69 |
| Appendix 2:   Kepler's Laws | 71 |
| Appendix 3:   Leap Second | 72 |
| Appendix 4:   Vulcan | 73 |
| Appendix 5:   Life on Mars | 74 |
| Appendix 6:   Corkscrew Path of Light | 75 |
| Appendix 7:   Birth of Planets | 77 |
| Glossary | 78 |
| Suggested Reading | 82 |
| Thank You | 84 |
| About Maureen | 86 |

# ILLUSTRATIONS

|  | Page |
|---|---|
| Lunar Polka | 15 |
| Light Bent by Gravity | 38 |
| Energy Converted to Matter | 44 |
| Creation | 60 |

# Dedication

I dedicate this book to Brendan Berg, son of Maureen, who inspired me to put this book together while I awaited his imminent arrival into this universe.

Dear Maureen,

Your questions brought on by your studies in Earth Science have started a dialogue between us that has reached far into the study of the universe--cosmology. That your studies in Earth Science should bring out questions deep into cosmology is a credit to your teacher.

Your questions and the answers would fill a book. Some of the answers your questions brought out go beyond presently accepted knowledge, and my concepts differ from presently accepted knowledge. For these reasons I think you should have the answers written out for you so that you can scrutinize them, tear them apart, question them, prove them, disprove them, and accept or reject them.

Therefore I will try to write you one short letter each day, following a logical outline until I have recorded for you the questions and the answers.

<u>Where does all the energy from the sun and stars go?</u>

What force is causing the galaxies to fly apart?

We're told that some of the galaxies are receding from us at speeds approaching the speed of light. What happens as they or we reach the speed of light?

Most people nowadays believe the universe started with a Big Bang! All the matter of the universe was collected in one ball, which exploded, sending the matter off in all directions to form galaxies, which are moving away from us and from one another at an accelerating rate. That is, the universe is expanding.

A few others theorize that the universe is in a steady state. When asked how it remains in a steady state while expanding--that is, what fills the voids left by expansion--they say that hydrogen is being created continuously.

<u>Creation?</u> Wow! <u>How?</u>

What are quasars?
Where might comets come from?
Asteroids?
How are galaxies formed?

Stars?
Planets?
Where does interstellar gas come from?
Cosmic rays?

These questions are listed not in the order in which they were asked, nor in the order in which they will be answered. But we will make an honest effort to answer each of them.

I think you knew when you asked the questions that some of them have not been answered to your satisfaction before.

> Bless this food
> and us
> by Your presence, Lord.
> Make us worthy instruments
> in Your continuing work of creation,
> for the good of man
> and the glory of God.
> Amen.

Grace composed by the author for dinner meetings of the Professional Engineers Society of Union County, New Jersey.

# OUR EARTH

Our Earth, as you learned in Earth Science, is an imperfect sphere of molten rock and metals, covered by a crust interspersed with oceans and covered by an atmosphere in which we live. The imperfect sphere is slightly flattened at the poles and bulges around the equator. The bulge is said to be caused by centrifugal force as the earth spins.

The earth is 8,000 miles in diameter, 25,000 miles in circumference and contains virtually all the known elements. It has an atmosphere that is 21% oxygen, 79% nitrogen plus traces of such elements as hydrogen, argon and neon. The interior of the earth is a molten magma containing practically all the known elements The crust is broken by "zits" that we call volcanoes from which the molten interior bursts forth from time to time, in sometimes catastrophic eruptions. The bottom of the sea contains not only volcanoes but also long cracks from which molten magma flows out, causing warm streams such as the Japan Current. The bottom of the Pacific Ocean is slashed with rifts, and the ocean is circled by volcanic lands.

The surface crust is broken up into what we call tectonic plates.

The plates form our continents. The plates are shifting to equalize unbalanced forces. Boundaries of the plates are called faults, such as the San Andreas Fault in California. As the plates move against one another the shifting and grinding and shingling of one against the other cause earthquakes, which occur mainly along the faults.

With all its faults, this is the earth we live on.

From a map of the world it is easy to see how the continents once fit together into what is called Panagaea. Panagaea covered only about one third of the surface of the earth. Did the crust once cover the entire planet? And, if so, where has the missing portion of the crust gone?

Where did this world come from? Some say it was flying around loose in the universe and was captured by the gravity of the sun. But where did it come <u>from</u>? A large glob of molten minerals containing all the known elements. Elements are believed to form within the tremendous pressure and high temperatures deep inside the molten centers of stars like our sun. Should we look at distant stars to find the origin of our molten glob? If it came from such a distant source, it would have been traveling for millions of years. Wouldn't it be congealed more than it is?

Or, more logically, did it come from the sun?

# THE MOON

The Moon is our most intimate celestial neighbor. It circles the earth about every twenty-four and three-quarter hours. Because the revolution does not take exactly twenty-four hours, we see the Moon in a different position relative to the sun each day. When the Moon passes between us and the sun, we see a New Moon. If the Moon blocks all or part of the sun, we have a solar eclipse. When the earth is between the Moon and the sun, we see a Full Moon. And if the shadow of the earth falls on the Moon, we have a lunar eclipse.

When I was a boy, we were still in a somewhat agricultural society, and every calendar showed the dates of the four main phases of the Moon. It was important because the Full Moon made it possible to work longer than the normal dawn-to-dusk day; this was especially important during the heavy-work-load periods of planting and harvest times. The "Harvest Moon" was acclaimed in song and dance.

# ORIGIN OF THE MOON

Where did the moon come from? A roving hulk that was captured by the gravity of the earth? Chancy.

Condensed from a cloud of gases? Our astronauts brought back rocks from the moon, so we know now that the moon consists of the same mix of matter that the earth does. And we're not accustomed to thinking of iron and the like blowing around in clouds.

Some say the marks on the surface of the moon indicate flying objects that have collided with the moon. When I hear this, I expect to see marks that are skid marks of near misses, not the perfectly round bulls-eyes most of the marks on the moon appear to be with their characteristic cusp in the center.

When I first made this observation, I was carried back to the time when I was a little boy and my nose barely reached above the coal stove as I impatiently watched chocolate pudding cooking. To renew my recollection, I cooked up a batch of chocolate pudding. Lo and behold, as bubbles of steam raised bubbles of chocolate pudding that then broke, there was the same circular remnant, with the cusp in the center, that I had seen on the pictures of the moon. How did such marks get there? If the moon were a molten glob of minerals, then as it cooled, its surface would cool first and congeal into a skin; and as bubbles of gas from the interior burst through the skin they would leave the same marks that I had seen on the surface of the pudding. In the beginning, the bubbles of escaping gas would be large and leave the large circles that appear to be old on the moon. As time went on and more and more of the gas expulsions escaped, the later bubbles would be smaller, making smaller circles. Just so, on the surface of the moon we see very large circles that appear old, with smaller circles within them. (We have, in fact, recently seen a photograph of volcanic activity on a moon, not on our Moon, but on Io, one of the moons of Jupiter).

Where, then, did the molten glob come from? Just as we proposed that our earth escaped or was expelled from the sun, let us propose that the moon escaped or was expelled from the earth.

Picture the earth, a molten glob, cooling and skinning over, the skin crusting, the crust shrinking, causing pressure on its molten mantle, the molten stuff held in by the toughening skin until sufficient pressure built up to burst out, the earth going through a period of high rotational velocity as it passes around a tight turn about one focus of its elliptical orbit, a large ball of the magma--one fiftieth of the earth--being ejaculated, carrying a third of the earth's skin with it, leaving the surface of the earth scarred, tearing a deep hole out of the earth, leaving a large cusp at the point of separation.

The earth was left in a mess. One third of its skin was missing, leaving the rest of the skin to break up and move around to equalize its unbalance--the tectonic plates; a tear deep into the magma of the earth--the Pacific rift; in the center of the rift the deep cusp at the point of separation-- the Hawaiian Islands.

We know that the earth has not always revolved placidly at 365.3 days per year. Fossils of sea shells have been found with 385 day lines per year ring. If the present rate of spin, 365.3 rotations a year, causes the earth to bulge at the equator, would 385 days in a year be a sufficient velocity to cause the earth to tear apart so? Or did it require a combination of accelerations? The combination could include the symphonic motions of the earth's orbit, the sun's orbit, and an elliptical speeding up of the local group of stars.

# THE PLANETS

Imagination is more important than knowledge.
                              Albert Einstein

Let's depict very briefly our Earth's companions as we know them.

Our Earth, eight other known planets, and a belt of asteroids all orbit about the sun. Thirty known moons orbit their planets. The planets with their moons orbit the sun all in one plane, very nearly.

Our sun's family of planets with their moons is a very orderly family, each unit balanced in its orbit by centrifugal force and gravity.

Table 1 (page 10) shows the ten planets and their distances from the sun. These distances were measured accurately by Bode, and Titus discovered that these distances are in a numerical proportion as shown in the equation:

$$a_n = 0.4 + 0.3(2)^{n-2}$$

n is the number of the planet from the sun: n = 1 for Mercury, 2 for Venus, 3 for Earth, etc.

The distances of the planets from the sun are shown in Table 1 on Page 13.

# SPACING OF OUR PLANETS

If we were to drop steel balls off a bridge at regular intervals of, say one second, and took a picture of them as we rolled the tenth ball off the bridge, we would find the balls spaced from the bridge at distances represented by the geometric progression: 1, 4, 9, 16, 25, 36, 49, 64, 81. This, Newton told us, is because the falling object is accelerated by a constant force, gravity, resulting in a constant acceleration, and his statement is embodied in the formula:

$$s = gt^2/2$$

where     $s$ = distance fallen
          $g$ = rate of acceleration due to the force of gravity
          $t$ = time

\*\*\*

The sun spun off one planet each time it passed through a certain point in its orbit of the galaxy. Each planet was acted upon by an outward, centrifugal force exceeding the inward, centripetal force of attraction to the sun. By the time nine such planets had been released they should be found to be spaced according to a geometric progression.

Meanwhile, some refinement of our technology--which has recently measured the rate of recession of the Moon--should verify the rate of recession of the planets, predicted here. A simple application of radar should give the rate of recession of the moon from earth directly. A direct correlation to the rate of recession of the galaxies will be found to exist. It should also give us Hubble's Constant, the rate of acceleration by which the galaxies are receding from us, the rate of expansion of the universe.

Note: One of our astronauts on his visit to the moon carefully placed a reflector facing earth to enable us to measure accurately the distance and rate of recession of the moon.

| PLANET | AVERAGE DISTANCE FROM CENTER OF SUN IN AU* (T) | LENGTH OF YEAR IN EARTH YEARS (R) | $\dfrac{R^3}{T^2}$ A.U. DAYS |
|---|---|---|---|
| MERCURY | 0.389 | 88 DAYS | $7.64 \times 10^{-6}$ |
| VENUS | 0.724 | 224.7 DAYS | $7.52 \times 10^{-6}$ |
| EARTH | 1 | 365.3 DAYS | $7.50 \times 10^{-6}$ |
| MARS | 1.524 | 687 DAYS | $7.50 \times 10^{-6}$ |
| ASTEROID BELT | | | |
| JUPITER | 5.2 | 11.9 YEARS | $7.49 \times 10^{-6}$ |
| SATURN | 9.5 | 29.5 YEARS | $7.43 \times 10^{-6}$ |
| URANUS | 19.4 | 84 YEARS | $7.77 \times 10^{-6}$ |
| NEPTUNE | 30 | 164.8 YEARS | $7.76 \times 10^{-6}$ |
| PLUTO | 39.8 | 248.4 YEARS | $7.77 \times 10^{-6}$ |

\* A.U. ASTRONOMICAL UNIT = DISTANCE OF EARTH FROM SUN (AVERAGE).

NOTE: THIS TABLE WAS ORIGINALLY WRITTEN BY KEPLER FOR THE FIRST SIX PLANETS; THEY WERE THE ONLY PLANETS KNOWN IN HIS TIME. COMPUTATIONS FOR THE REMAINING THREE PLANETS HAVE BEEN ADDED BY THE AUTHOR. NOTE, IN THE LAST COLUMN, THE CLOSE CORRELATION TO KEPLER'S WORK.

These distances are the same as they would be if the planets started out by being ejected by the sun one at a time, at regular intervals and accelerated away from the sun each at the same rate of acceleration.

Here, then, we have nine planets, made of elements of the kinds that are formed in the high temperature and pressure of the center of the sun, and spaced as though they had been ejaculated from the sun at regular intervals. What could be more logical than to propose that the planets were ejaculated from the sun at regular intervals?

What intervals? We have Kepler's assurance that the planets travel in elliptical orbits. We know that, traveling in elliptical orbits, bodies increase their speed of rotation about their polar axis and their speed of travel in orbit as they travel through the tight turn of their ellipse (See Kepler's law of equal areas in the Appendix.) It is at that point in the ellipse that the centrifugal force on the body is greatest. Can we then imagine that, under these effects of extreme centrifugal force, at the tight turn of the ellipse, the sun ejects a body that then congeals and becomes one of its planets?

Following Einstein's suggestion then, and letting our imagination work, let us propose that the planets started as bodies of molten matter being ejaculated from the sun at regular intervals, the regular interval being the time when the sun passed through the tight turn of its elliptical orbit about its focus.

Separation of a planet from the sun is a cataclysmic affair. The sun bursts apart into two joined pieces. It isn't clear which of the two is the larger. It appears to be a tug of war in which one gradually gains over the other in a shower of flaming molten matter. The larger settles into the role of the sun, the smaller settles into the role of the new planet. Some globs escape, do not join one or the other, but travel in an ellipse about the pair, with the sun as one focus and the planet as the other focus. As the pair separates, these chunks become comets. It is known that some comets travel as far as Jupiter. Others are believed to travel for thousands of years. I believe their outer foci will be found to be the locations of distant, as yet unknown planets of our sun.

The flaming, tearing pair looks, from a distance, like a dual star. In fact, many of the stars in the universe appear to be dual stars: stars giving birth to their planets.

# SOLAR SYSTEM

We know that the planets rotate about the sun, each constrained in its elliptical orbit by the balance of the sun's attraction and the centrifugal force caused by rotation of the system.

The farther from the sun a planet is, the slower its rotational speed, and the longer its "year"--the time the planet takes to make a complete circuit of the sun--as shown in Table 1.

Each planet exerts a force on each other planet. Neptune was discovered when an astronomer discovered a deflection in Uranus' orbit. The astronomer calculated the mass, distance and direction of a body required to cause the deflection, directed his telescope at the spot and found the eighth planet, which he named Neptune in keeping with the practice established by the Greeks of naming the planets for Greek gods. The ninth planet was found in the same way, and was called Pluto.

All the planets orbit the sun in the same direction.

All the planets except Pluto travel in very nearly the same plane, which is called the ecliptic, the plane of eclipses.

Between the planets Mars and Jupiter a circle of asteroids, like a planet broken into chunks, travel in their own orbit.

Besides Earth, five of the planets have moons, and more moons are being discovered.

Saturn is circled by bands of rubble, the rings of Saturn, in addition to nine moons.

We seem fairly satisfied that rotation of the solar system and the balance of centrifugal forces (due to rotation) against centripetal forces (due to the attraction that the Sun exercises on its planets) keep the various bodies in orderly orbits and prevent them from bumping into one another.

We have seen by the preceding equation and table that the planets are spaced from the sun and from one another in a very orderly formula.

But what about the stars? What keeps the "sky from falling"?

Let us look farther out into space.

Does it strike you that there are some similarities between the opposing forces of gravity and centrifugal force?

For example, centrifugal force is the <u>only</u> force that effectively opposes gravity.

And: the attraction of gravity is <u>inversely</u> <u>proportional</u> to the square of the distance between the bodies, while centrifugal force is <u>directly</u> <u>proportional</u> to the square of the distance between the bodies.

Also: the attraction of gravity is directly proportional to the mass of the body, and centrifugal force is also directly proportional to the mass of the body.

In equilibrium, as in a satellite in fixed orbit, centrifugal force is exactly equal to the attraction of gravity.

It would seem that gravity might somehow be the opposite of centrifugal force, and have something to do with rotation, or "inverse rotation" to coin a phrase to express the incipient thought.

Like the grandfather's flying coattails--see the illustration on the next page--the tide on the far side of the world is caused by centrifugal force. Isn't it surprising that centrifugal force can raise the level of the seas <u>four feet?</u>

# CENTRIFUGAL FORCE

> "The Sky is Falling,
> The Sky is Falling"
> cried Chicken Little.

What keeps the Moon from falling down on the Earth? Centrifugal force, of course. The Moon is kept in its orbit by the balance between two forces: centrifugal force, due to its rotation about the Earth, and the attraction of gravity between the Moon and the Earth.

Actually, we believe the two forces are not perfectly balanced, and the Moon is moving farther from the Earth by the tiny distance of about three centimeters each year.

And the Moon is not rotating about the Earth, but the Moon and Earth are rotating about their common center of mass, like a portly grandfather dancing a lively polka with his petite granddaughter.

# ATTRACTION

We are told that every body in the universe is attracted to every other body in the universe. We are also told that the force of attraction between them is directly proportional to the product of the masses of the two bodies, and inversely proportional to the square of the distance between them.

Of all the heavenly bodies, the Moon, being closest, has the greatest attraction for Earth.

Because of the effect of the square of the distance, the massive Sun, being farther away, has less attraction for Earth than the smaller Moon has.

The seas of the Earth respond to the attraction of the Moon and give us the rise and fall of the tides.

When I was a boy, Elizabeth Taylor, the child movie star, had a swimming pool, and so did dozens, perhaps hundreds of people. The rest of us had oceans, bays, and rivers. We boys had Newark Bay, and we watched the newspapers for the time of high tide. High tide occured <u>twice</u> a day, and this always puzzled me, because the moon passes overhead only <u>once</u> a day. I could understand the water rising four feet toward the moon, but what caused an equal tide on the opposite side of the earth?

Do you know? Don't turn the page until you've thought about it.

# MORE ON ATTRACTION

While the seas facing the moon are attracted to the moon, causing high tide, an equally high tide is thrown up on the opposite side of the earth by centrifugal force.

The earth is attracted by not only the Moon but also by the Sun and all the planets, and by every star in the universe.

After the moon, the Sun exerts the next most noticeable effect of attraction on the earth. The attraction of the Sun affects the tides noticeably. When the Sun and the Moon are pulling in the same direction, we have the highest tides; and the Earth, swung like a skater on the end of a line of skaters playing snap-the-whip, throws up an equally high tide on its outward side due to <u>centrifugal force</u>.

While the force of attraction is most noticeable in the tides, the attraction is exerted also on the crust of the Earth and on the Earth's fluid center.

# ATTRACTION EXISTS BEYOND THE SOLAR SYSTEM

We have stated that the Earth wobbles in its orbit, the wobble being caused as Earth and Moon rotate about their common center of mass.

Binary stars are pairs of stars traveling <u>a deux</u> like our Earth and Moon. The existence of binary stars tells us that mutual attraction exists beyond the solar system. Indeed, binary stars in which one of the pair is invisible, have been discovered when astronomers found the visible member of the pair wobbling in its orbit.

<u>This was proof that attraction exists beyond the solar system.</u>

# ROTATIONAL ENERGY

To prepare ourselves to understand some of the phenomena that can be observed taking place over millions of light-years in the universe, let us examine the principle involved in Rotational Energy. There are some simple experiments that we can perform in a lavatory laboratory.

First, fill a washbasin sink with water. Pull the plug and let the water in the bowl run down the drain. Does the water in the bowl rotate? If the bowl is fairly round and the stopper is an old-fashioned rubber stopper there should be very little difficulty in observing a swirling motion. If the stopper is a metal gadget operated by a lever you might remove the metal stopper, but it's less trouble to flush the toilet and watch the water swirl. I used to think the engineers who devised that swirling, scouring action were very ingenious, until I discovered they it had come easy to them.

If neither of your plumbing fixtures produces the swirl, you may need to exercise some ingenuity in finding a sink of some sort that does. It is important, before reading on, that you observe the swirl of the water as it is released to go down the drain.

In which direction does the water swirl, clockwise or counterclockwise? Why? Think about it before reading further.

# CONSERVATION OF ROTATIONAL ENERGY

If you have thought about why water spins as it goes down the drain, now let's think it out together.

Take your world globe into a darkened room. Turn on a lamp and let it represent the sun. Now rotate the globe so that dawn moves across the United States from east to west, from New York to San Francisco.

Keep the earth spinning and view the North Pole. In which direction is it spinning, clockwise or counterclockwise?

You observe that the world is spinning counterclockwise when viewed from above the North Pole.

Keep the globe spinning in the same direction and, from below, view the South Pole. The South Pole is spinning in a clockwise direction, isn't it?

And you have learned in Earth Science that we have cyclones in the northern hemisphere and anti-cyclones in the southern hemisphere.

You are fortunately about midway between the equator and the North Pole.

Let us also say here that the rotational energy of an object is composed of two parts: its rotational energy about its own polar axis and its rotational energy about the center of its orbit. As an example, the rotational energy of the Earth in the solar system is composed of its rotational energy about its own polar axis plus its rotational energy in its orbit about the Sun.

# MORE ON CONSERVATION OF ROTATIONAL ENERGY

Does this give you some hint as to why the water swirls so rapidly when you let it down the drain? Let us call the phenomenon a manifestation of the Principle of Conservation of Rotational Energy. The Principle should be observable in many phenomena in a rotating universe. Perhaps it can give us a test or two to prove whether or not the universe itself is rotating!

Let's have a clear statement of the Principle of Conservation of Rotational Energy: The total Rotational Energy of an object consists of two parts: (1) the rotational energy of the object about its own internal axis of revolution and (2) the rotational energy of the object about one focus of its orbit; and inertia causes the object to resist any change in total rotational energy.

Remember that the earth's orbit is an ellipse, nearly circular, but not quite. Therefore, the earth will be closer to its focus at some times, and farther away at others. This distance could change its revolutionary speed about the focus, except that the earth speeds up its revolutionary speed about its focus when it is closer to its focus, and slows down when it is farther away. Refer to Kepler's Law of equal areas (Appendix).

Any loss of rotational energy about the focus of its orbit must be made up by an increase in the rotational energy about the object's own internal axis of revolution so that the total rotational energy of the object remains unchanged.

# ROTATION AND ATTRACTION

It is said that a pair of scientists set a gyroscope rotating in a suitcase in a Paris railroad station and called a porter to carry the suitcase. When the porter picked up the suitcase it jerked southward. When he tried to pull it toward him it turned West, he dropped the suitcase and fled North, midst the howls of laughter of the two tricksters.

As you probably know, a gyroscope makes a very dependable compass; once set rotating pointing North, and powered to keep rotating, it can stay pointing North through all but the worst battering a ship might encounter in a storm at sea.

Concerning this subject, do the following experiments:

Make a simple pendulum of a weight and a sufficient length of twine such that, when suspended from the ceiling, the weight hangs within an inch or less of a table top. Mark the spot directly under the weight. Now pull the weight a foot to one side and let it swing freely. Observe the path of the weight. Does the path of the weight cross the spot you marked?

Repeat the experiment, starting from different directions from the spot. One time, release the weight and, with a chalk, mark the path of the weight through eight or more swings. Now untie the plumbob and in its place tie a toy gyroscope. Set the gyroscope spinning. Release it carefully as before. Does it pass over the marked central spot? Why not?

With a chalk, mark the path of the gyroscope pendulum through eight swings. Describe the path of the gyroscope pendulum. Really <u>do</u> these experiments before reading further. If you do, you will, later in these experiments, experience the superb joy of <u>discovering</u> for yourself a secret of the universe that Einstein and Newton sought.

# ELLIPTICAL GALAXIES AS EVIDENCE OF ROTATING UNIVERSE

In these simple letters from your father to his daughter, I have not made a point of identifying the great men who made the major discoveries in astronomy and cosmology. It is difficult however, to discuss elliptical orbits without mentioning Johannes Kepler. Two hundred years ago Kepler discovered that all planets in the solar system travel in elliptical orbits around the sun. Kepler told us a great deal about these elliptical orbits, but he did not tell us <u>WHY</u> the planets travel in elliptical orbits.

If you performed the simple laboratory experiment suggested in one of my letters of suspending a toy gyroscope and letting it swing as a pendulum, you discovered that the spinning gyroscope is attracted by gravity not directly, as a plumbob pendulum, but indirectly, so that the spinning gyroscope oscillates in an <u>elliptical</u> path.

It is my intent not to engage in a mathematical discourse beyond your scholastic grade. I fervently hope that your interest will continue and that we may some day develop some of the appropriate mathematics together. In a year or two you should have the background to appreciate any of several good books of the mathematics of the gyroscope. It will enhance your understanding of the reasons why A ROTATING BODY IN A ROTATING FIELD TRAVELS IN AN ELLIPTICAL PATH. If Kepler had the information we have today, I feel he would have written this as his fourth law.

A century and a half after Kepler we discovered other universes: "island universes", galaxies. Astronomers have observed that these galaxies are elliptical in shape, as a general rule.

Why do the galaxies form in elliptical shape? <u>BECAUSE A ROTATING BODY IN A ROTATING FIELD TRAVELS IN AN ELLIPTICAL ORBIT.</u>

<u>The rotating field is the rotating universe.</u>

# ROTATING UNIVERSE

Let us develop a picture of our rotating universe.

Any rotating universe MUST be finite. There seems to be just no way to picture infinity spinning.

Let us accept for the time being that its peripheral speed at its outer surface at its equator can not exceed the speed of light. Our observation tells us that galaxies are receding from us at speeds approaching the speed of light. If we accept the premise that matter cannot travel at the speed of light we must conclude that the matter must change direction or form before it reaches the speed of light. Therefore we reach the conclusion that no matter exists at the outer extremity described above, that is, at the surface whose peripheral speed is the speed of light.

A little thought on the subject brings the picture of a rotating spherical universe in which most of the matter is located in or near a disc in the equatorial plane, much as the matter of the Andromeda Galaxy is concentrated in a swirling disc.

This is a brief sketch of our uni-verse: rotating unit.

# THE PATH OF A COMET

Have you ever wondered why Halley's Comet can be attracted by the sun so strongly that it returns from so great a distance after eighty years, yet doesn't plunge right into the sun, but narrowly misses it? Some astronomers would think you ignorant to ask, and would answer "Why, that's its orbit!" Don't you believe it! Since Halley's Comet's last visit, the sun has been traveling at twelve miles a second for eighty years.

Stop reading, close this book, go get a snack, and while you're eating it, think about why Halley's Comet and tens of thousands of other comets just miss the sun time after time after time.

***

If you've figured out why comets miss the sun, let's prove your answer, and if you haven't, let's discover it:

Repeat the pendulum and gyroscope experiment while thinking about the comet and its path.

Hang the ordinary pendulum still and mark the center spot.

Swing the ordinary pendulum and note that its path passes through the central spot. Replace the pendulum weight with the toy gyroscope, set the gyroscope spinning, and let it swing freely.

Note that the rotating gyroscope, while attracted by gravity, does not pass over the central spot, but grazes it in a series of ellipses.

Why?

Now think out why the comets miss the sun.

Ponder this question while rereading the preceding half-dozen letters.

# WHERE DOES A COMET GET ITS SPIN?

We discovered that a comet misses the sun because a rapidly spinning object is attracted, not directly, but askew.

Where does a comet get its spin?

To examine this question, let us start with the comet at its aphelion, its farthest point, millions of miles from the sun, not spinning about its own center, but a part of the rotating solar system, and as such, having a great deal of angular momentum about its distant focal point, the sun. In this quiescent state, the comet is a peaceful, attractive companion to the neighboring particles of frozen nitrogen, hydrogen, dust particles, and other cobwebs of the cosmos, and it attracts and collects them.

From its aphelion, under the unrelenting attraction of the sun and planets, the comet commences its approach to the center of gravitational attraction of the solar system. Picture the attractive influence of the solar system as a rotating gel in which the comet is traveling.

As the comet approaches the sun, its radius of gyration about the sun is shortening, and so its rotational energy about the sun is decreasing. But we have found that rotational energy resists dissipation and tends to be conserved. How can the rotational energy of the comet conserve itself?

<u>By the comet rotating about its own axis.</u>

And so the comet takes on an increasingly more rapid rotation as it approaches the sun. This is the rotation that gives it a gyroscopic action, and causes the comet to graze the sun and not plummet directly into it. The rapid rotation of the comet in proximity to the sun casts off some of the frosty covering of frozen solid hydrogen and other matter, which then form the comet's tail.

Centrifugal force resulting from the comet's tight turn about the focus of its orbit overcomes the centrifugal force of the comet's rotation about its internal axis, and causes the comet's tail to point away from the sun as the comet races through the tight curve of its perihelion, its closest approach around the Sun.

# GALAXIES

When I was about your age we didn't have television, and newspapers brought to our wondering eyes the news of the world including "all the news that's fit to print" (and some that wasn't).

One story that was so newsworthy as to command the front page of the Sunday Supplement in Rotogravure for several Sundays was the <u>discovery of other universes than our own!</u> Telescopes had become good enough to see that certain blurry stars that had, until then, been called "nebulae", were not single stars at all, but disc-like groups of thousands of millions of stars. There were many such groups, each isolated by itself in the vastness of space, and they were referred to as "island universes", two words that, seen together, could make a pair of eyes stop reading, go glassy, and see more than the whole newspaper could possibly convey in words. <u>Island Universes, Galaxies</u>.

We were told that our Sun was one undistinguished star of a thousand million or so in our own Galaxy.

# A GALAXY OF GALAXIES

The first galaxy to be known was a blurry spot which had been just barely visible to the unaided eye all along, but now became explorable by the latest telescope.

Stargazers, in order to be able to point out to one another various locations in the sky, long ago had named the groups of stars that appeared as crosses, dippers and particularly, familiar animals. They called these groups of stars "constellations".

The blur that now received the astronomers attention appeared in the constellation called "Andromeda", and so it became known as the "Andromeda Galaxy".

Shortly after Andromeda, other fuzzy lights, "nebulae", were investigated and found to be galaxies! Thousands of them! Millions! Each containing thousands of millions of stars. Old stars, new stars, red stars, blue stars, twinkling stars, and stars that are spinning in pairs.

Astronomers got a better picture of our own Galaxy by studying the newly-found distant galaxies, which displayed themselves at all angles. Galaxies were seen to be pancake-shaped with a bright bulge in the middle, somewhat like a fried egg over lightly, the yoke bulging in the center. Some had arms like pinwheels. Most were elliptical in plan view. Detailed study of the pancake, and of the tips of the pinwheels, showed that the galaxies were spinning.

The great milky streak across our sky had been known for hundreds of years to consist of thousands of millions of stars, and had been considered to be _the_ universe. Many had wondered why the majority of distant stars were gathered into one path across our sky. Now we knew that the "Milky Way" was our view of our own pancake Galaxy from our position within it. And so our Galaxy became known as the "Milky Way Galaxy".

We who were growing big enough to read the Sunday Supplement, and too big to spend our weekly allowance on "a penny's worth of these and a penny's worth of those", showed our sophistication by spending our whole nickel on one candy bar which was just beginning to appear in the glass-enclosed candy counters alongside Hershey Bars, and was called "Milky Way" as the candy maker's way of celebrating the new discovery and capitalizing on it.

# OUR ROTATING GALAXY

What keeps stars in our Galaxy from coming together under the influence of their mutual attraction?

We have seen that the Moon keeps its distance because the Earth-Moon system rotates. Centrifugal force overcomes gravitational attraction.

We have seen that the planets keep to their orderly orbits around the Sun because the solar system rotates. Centrifugal force balances gravitational attraction.

We see that our Galaxy rotates. Although we do not know the orbits of the stars in the Galaxy as we know the precise orbits of the planets in the solar system, we should find it fairly evident and acceptable that centrifugal force due to rotation of the Galaxy can balance the attractive forces acting between stars, to keep them apart.

<u>Centrifugal force and only centrifugal force balances the attraction of gravity.</u>

# OUR ROTATING GALAXY

Q. What keeps the Moon from falling on the earth?
A: Centrifugal force. The pair rotates.

Q. What keeps the planets from falling into the sun?
A. Centrifugal force. The solar system rotates.

Q. What keeps a galaxy from collapsing into itself under the attraction of its own gravity?
A. Centrifugal force. The galaxy rotates.

## QUIZ

Q. What keeps the universe of galaxies from collapsing into itself under the attraction of its own gravity?
A. _____ _____. \_\_\_ _____ _____.

CHECK YOUR ANSWER WITH THE ANSWER ON THE NEXT PAGE.

A.  Centrifugal force. The universe rotates.

   Now you know something Einstein didn't know.

# WHAT KEEPS GALAXIES APART?

What keeps galaxies apart? What force opposes the mutual attraction between galaxies? What prevents all the galaxies from being drawn together?

We have learned that Moon and Earth are held apart by centrifugal force resulting from rotation. The planets are attracted to the sun but resist the attraction by rotating about the sun. The stars of a galaxy resist attraction by the rotation of their "island universe".

What more natural answer to the question "What keeps galaxies apart?" than <u>centrifugal force due to ROTATION OF THE UNIVERSE keeps galaxies apart.</u>

Let us propose the hypothesis of a <u>rotating universe</u> and examine how a rotating universe suits the many phenomena, some explained, some unexplained, and some very unsatisfactorily explained until now.

## uni-verse: rotating unit

And so we propose that the observable galaxy of galaxies, the thousands of millions of galaxies that we observe with our visual telescopes and with our radio-telescopes constitute one rotating unit.

To avoid too many "would's" and "should's" we will avoid the subjunctive mood and use the indicative mood. It should be observed, however, that the proposition of the rotating universe is a hypothesis.

We shall examine how the proposition of a rotating universe may be supported by observable features such as the shape and rotation of the galaxies.

We shall look to the rotating universe for explanations of quasars and of the apparent expansion and renewal of the universe.

We shall ask of the hypothesis of the rotating universe: rotating relative to what?

What happens to the prodigious amounts of energy that are expended by conversion of matter to energy in the stars?

How does the expanding universe renew itself?

Explain anti-matter.

What causes gravity?!

Through all this ambitious dissertation, Maureen, please bear in mind that it was brought on by your questions, and keep this disclaimer before you: it is only a hypothesis, to be questioned, rejected, accepted, disbelieved. If these writings are to serve a worthwhile purpose it is not to answer your questions but to instigate in your mind more probing questions, to give you the audacity to ask your questions, and the courage to defend your answers.

# WHAT ARE QUASARS?

Radio telescopes have discovered powerful sources of radio waves emanating from pinpoint locations hundreds of thousands of light years away, at the very outer frontiers of the universe. The output of each of these sources is more powerful than anything seen or known in the nearby universe.

Such sources of radioactivity are called "quasi-stellar radioactive sources", or quasars for short.

A nova is an exploding star. A quasar is as powerful as a galaxy of exploding super-novas.

In our rotating universe let us propose that galaxies are born near the center of the universe and move outward, continually gaining velocity until they attain a critical velocity, at which they explode in a shower of power and radioactivity: quasars.

Scientists are finding quasars in all directions. Probing beyond the quasars they find ..... nothing.

# ENERGY IS ATTRACTED BY GRAVITY

We talked about objects being attracted by other objects. Did you know that rays of energy are attracted by objects?

Young Albert Einstein in his Theory of Relativity predicted that light waves would be attracted by gravity. Scientists were anxious to prove or disprove this, so they devised a test.

At a certain time the Moon would blot out the Sun just as a star was about to be blocked by the Sun. Under these conditions, with the Moon completely blocking the Sun, the day would become completely dark and the star could be seen. Astronomers would track the star to find out whether its apparent path was deviated by deflection of the light rays from the star passing through the strong gravitational field of the Sun. Their observations showed Dr. Einstein to be wrong! When they told him so, he, without looking up from his work or missing a puff on his pipe, told them they were careless in their observation.

See the diagram on the following page.

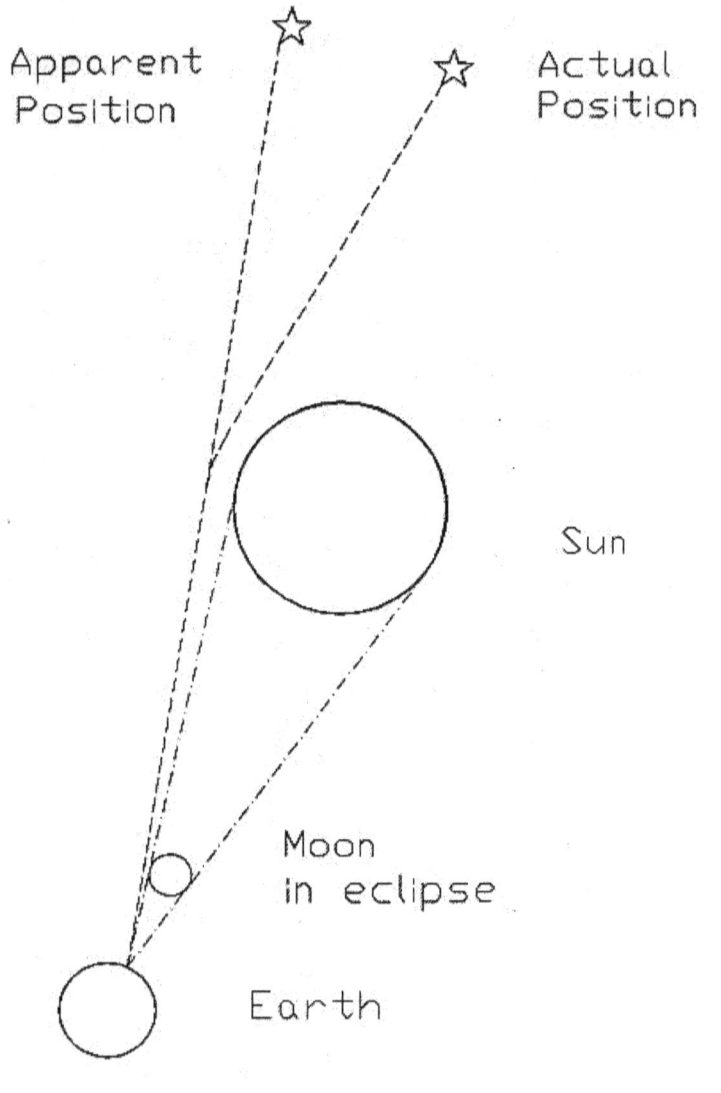

Light Bent by Gravity

So one astronomer set up his telescope in the desert. He photographed the path of the star one year before the next eclipse. Then, he protected his telescope from wind and sand and waited a year for the eclipse. Sure enough, he found the rays were deflected so much in passing through the Sun's gravitational field that he could "see" the star long after it had passed behind the Sun. The leading astronomers of the world converged on the site and confirmed his observation. See the illustration on the facing page. Energy is indeed attracted by gravity.

# ENERGY WAVES ATTRACTED BY ENERGY WAVES

If the sun attracts a celestial body, does that body attract the sun?

Do you think that the ray of light that was attracted by the Sun attracted the Sun?

If two rays of light traveling right next to one another are both being attracted by the Sun and are both attracting the Sun, do you think they attract one another?

## MATTER TO ENERGY
## ENERGY TO MATTER

Matter is converted to energy in the Sun and the stars of the universe at a prodigious rate.

Is all that matter lost by the universe forever?

What happens to all the energy produced? Does it raise the temperature of the universe? Or is all that energy radiated out of the universe and lost forever?

***

It is unthinkable that the prodigious quantities of matter that are being converted to energy each microsecond of time should be forever lost to the universe.

It is more logical that <u>energy is being reconverted to matter at the same rate that matter is being converted to energy.</u>

BUT HOW?

# ENERGY BECOMES MATTER. ENERGY BECOMES MATTER AT THE SPEED OF LIGHT SQUARED.

$$E = mC^2$$

We're more familiar with it in the atom bomb: Matter becomes Energy.

But Einstein didn't write: $m = \dfrac{E}{C^2}$   "Matter becomes energy at the speed of light squared"

Yet that's what he accomplished in the atom bomb.

There's an anecdote that a junior high school boy had trouble with an algebra problem in his homework, and he sent the problem to Einstein, who scribbled off a quick solution and returned it to the boy, and it had a mistake. Even so, Einstein was too good a mathematician to use the equal sign if he meant the equation worked in only one direction; he would have used an arrow. We can be confident in interpreting Einstein's equation:

"Energy becomes matter at the speed of light squared"

Note: Speed of light <u>SQUARED</u>? That has puzzled me.

I picture first a vector representing the speed of light traveling North, and then that vector being translated Westward at the speed of light. Quite a rapidly expanding area.

As an alternative, picture an electron orbiting a nucleus at the speed of light while the nucleus and electron move in a straight line at the speed of light. This makes a more useful picture to me.

# HOW IS ENERGY CONVERTED TO MATTER?

It isn't easy to draw a defensible physical picture of an atom, but let's try.

Atomic theory tells us that the simplest atom, hydrogen, consists of an electron in its orbit about a nucleus. We draw such an atom like a solar system of a sun with a single planet. We generally picture the electron revolving about the nucleus in all directions, not a single plane, but this may be incorrect.

Separating the electron from the nucleus releases a certain amount of energy.

A great deal more energy, nuclear energy, resides in the nucleus. There is a great deal we have yet to learn about the nucleus, but we can picture it as small particles of energy in wave form, and picture the wave like the elastic band of an old-fashioned golf ball, wound and wound on itself. Where the energy of the nucleus is released, let us picture one of the rays of energy as a helix like a long spring, formed by untangling the elastic band into a long straight helical spring.

With this picture we see the problem of converting energy to matter as the problem of tangling the long helical spring back into ball form.

Let us consider a ray of light traveling in a great circle on the periphery of the universe, and being drawn inexorably closer and closer toward the center of the universe by the relentless pull of gravity.

As the ray of light approaches the center of the universe, closer with each orbit, it must obey the law of Conservation of Rotational Energy. On the outer periphery of the universe, traveling in a great circle at the speed of light, the ray of light had great rotational energy.

As it approaches the center how can it absorb more rotational energy?

Let us consider one wavelength of the ray of light.

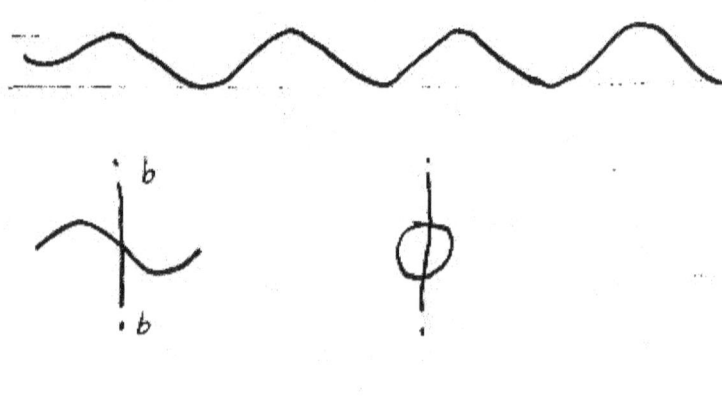

**CONVERSION OF MATTER TO ENERGY TO MATTER**

Obviously it cannot rotate about its present axis of rotation any faster, since it is rotating at the speed of light. But it might tumble about a transverse axis, depicted by b-b, in which case its configuration might be shown as in this sketch, while continuing to travel in its original direction.

Isn't this precisely the configuration we said represents matter?

If so, we have traced the transfiguration of one wavelength of light from <u>energy</u> <u>to</u> <u>matter.</u>

# MATTERGY

$$E = mc^2$$

We are familiar with "Matter becomes energy at the speed of light", as in the atomic bomb and the nuclear bomb.

Now we will learn: Energy becomes matter at the speed of light.

Matter and energy are interchangeable and the line separating them is the speed of light.

It seems necessary to coin a word to name this entity that can be either matter or energy: let's call it "Mattergy".

Mattergy is that entity that is energy at the speed of light, and matter below the speed of light.

# EXPANDING UNIVERSE

From measurements of the light from distant galaxies, we believe the universe is expanding. Light from distant galaxies is shifted toward the red end of the spectrum. This is believed to be caused by the galaxies moving away from us at high speeds, approaching the speed of light. The Doppler Effect. Galaxies in all directions are moving away from us. Hubble told us that the red shift indicates that the universe is expanding. We are looking for the Hubble Constant: the measure of the rate of expansion of the universe. The speed of movement of the galaxies is calculated by the change in frequency of the light shown by the red shift, and the distances of galaxies are estimated by the brightness of the galaxies. The farther away the galaxies are, the faster they are receding. Astronomers believe that the brightness of certain variable stars is proportionate to the period of their variation. Cepheid Variables. This gage is useful up to half the distances to the most distant galaxies, and from that point on we are dependent on the apparent brightness of the galaxies themselves. Differences in the two methods give us two figures for the Hubble Constant and they differ by 100%.

Must we depend on such distant objects? Or are the characteristics of the universe universal? Does the expansion of the universe start in the distant galaxies that are so far away and moving so fast that they show red shift? Or are the characteristics of the universe universal? Is the universe expanding right here in our own cosmic neighborhood? Are we to write on one page that our sun is a quite ordinary star, and, on another page that its solar system expands on its own set of rules, different from the remainder of the expanding universe?

The spacings of the planets indicate that they started by being ejaculated from the sun at regular intervals of time, and precise measurements of their distances indicate that they have been moving away from the sun at a constantly accelerating rate. Is the cause of their accelerating dispersion the same cause that is causing the universe to expand? If so, and if we can measure the rate of acceleration of the planets, have we not measured Hubble's Constant?

The moon, we have read, started by being ejaculated from the earth. The distance of the moon from the earth has been measured recently to be increasing by a small amount, a few centimeters a year.

The outward acceleration caused by centrifugal force must be greater than the inward acceleration caused by gravitational attraction. We ought to be able to measure the distance to the moon accurately. If we do, we can try to see whether or not that same acceleration applies to the motion of the planets. If so, Then: have we not measured Hubble's Constant in cm/yr/A.U. right here in our own neighborhood, rather than in kilometers per second for each megaparsec of distance from earth to distant galaxies?

# REGENERATING UNIVERSE

Observation tells us the galaxies are flying away from us, and the farther away from us, the faster those galaxies are receding, some reaching a large part of the speed of light.

In our rotating universe galaxies disintegrate as they approach the speed of light because, as we just discussed, matter turns to energy at that speed. These disintegrating galaxies give off the prodigious radiation which we observe and have quizzically called quasars, (quasi-stellar radiation sources).

No matter can leave our rotating universe, since its outer surface is traveling at the peripheral speed of light.

Our rotating, finite universe has a center of gravity, with all the pull of all the material of the universe.

The flying detritus of all the material of the exploding, disintegrating galaxies, although attracted to the gravimetric center of the universe, is propelled outward by the intractable centrifugal force caused by rotation of the universe.

So long as it is in the form of matter, and has mass, it is propelled outward by centrifugal force. After the matter has been thoroughly exploded into energy, and so has gotten rid of all its mass, it is no longer affected by centrifugal force and it can be attracted by gravity to move closer to the center of the universe.

The energy resulting from explosion of matter in the quasars at the outer limits of the universe is attracted by the gravity of the entire universe. Having no mass, the energy is not affected by centrifugal force. It is attracted to the center of the universe. Just back from the wild party in the garden of the quasars, our energy is highly stimulated; it arrives just as it crosses the line from energy to matter. It is matter in its simplest form: hydrogen, consisting of a nucleus and one electron. Because of this genealogy, hydrogen is the most plentiful material in the universe. The quasars are high production sources, and the mattergy arrives in raging jetstreams of white-hot hydrogen, to swirl and settle in places which we, on earth, seeing the commotion, call the birthplaces and nurseries of the stars. Here, the hydrogen atoms will be compressed en masse by their own gravity to form bodies like our sun. Within these infernos the hydrogen will be reheated and robbed of its electrons, which will then be available to re-form--within the tremendous pressure and temperature--into more complex atoms all up and down the atomic table. The stars so formed will be swirled out of their nursery and caught up into a family forming a new galaxy. A newly formed galaxy reaches a critical size, moves off center and, under the influence of centrifugal force caused by our rotating universe, starts its trip at ever-increasing speed outward toward.......... disintegration at the quasar fringe........And then recycling.

# AGGLOMERATION

In an earlier letter "How is Energy Converted to Matter" we traced the conversion of one wavelet of light from energy to matter. Conversion of one wavelet of light from energy to matter may be sufficient to establish the principle, but what goes on from there?

We can picture one ray of light emanating from a quasar --a galaxy exploding at the outer extremity of this universe-- the ray of light traveling a great circle on the outer surface of the universe. We have seen the diameter of this path reduced by the action of the gravitational attraction of the universe. Following the law of Conservation of Rotational Energy, we have proposed that, approaching closer to the center of rotation of the universe, the wavelet manifests its additional rotational energy by tumbling, catching its own tail, as it were, so that the quantum of energy travels in a spherical orbit, while the orbit continues to travel in a continually reducing great circle of the universe. What now?

Now, as this minisculum of mattergy in motion approaches closer to the center of rotation it must somehow assume additional rotational energy, without being exploded back into linear energy.

Our minisculum of mattergy could absorb additional rotational energy if it could join up with another minisculum, and the two rotate about a common center.

Presumably another wavelet from the original ray of light has converted to a minisculum of matter and is traveling along the same path in close proximity to the first. They join in pairs to rotate about their common center of mass, so they are able to satisfy the demand of the Law of Conservation of Rotational Energy.

# SPIN

To get an adequate picture of spin, let us reiterate what was said in an earlier chapter.

When you pull the plug in a sink or flush the toilet you notice that the water spins as it goes down.

Why does the water spin?

The water is about 3000 miles from the polar axis of the earth, and the earth is rotating. As a result, the water, though standing still in the sink, has much rotational energy about the polar axis of the earth.

When you open the drain, the water moves several inches closer to the axis of the earth before it disappears down the drain. Several inches closer to the axis of the earth, the water has less rotational energy about the axis.

But the principle of Conservation of Rotational Energy requires that the total rotational energy after the event must be equal to the total rotational energy before the event.

Since the water has less rotational energy about the axis of the Earth after the plug is pulled than it had before, what happened to the rotational energy it lost?

That difference in rotational energy manifests itself in the visible spin of the water.

With the principle of conservation of rotational energy in mind, let's take a look at a "particle" of energy, such as a photon of light, traveling along the "surface", the outer limit of our rotating universe. The particle is constantly attracted toward the center of the universe by gravity, and so is constrained to travel in a great circle along the universe's "surface". As it travels, eon upon eon, the photon-cum-particle is gradually attracted nearer the center. But near the center the particle will have less rotational energy. What happens to the rotational energy the particle lost?

Like the water going down the drain, the particle will develop spin.

When a particle has attained its full quota of spin, all it can hold, and it approaches closer to the center of the universe, losing still more rotational energy, how does this decrement of rotational energy manifest itself?

The only apparent outlet is for the particle to seek out another particle and the two spin about one another. Of course the two would have to be compatible regarding such features as electric charge, magnetic field and spin. And so we might have energy particles build up until we have an electron combine with a proton to form an atom.

Now, when this atom has absorbed its full quota of spin -- all it can take, and it approaches still closer to the center, what happens to its loss of rotational energy?

Right! It seeks another atom, is attracted to it, and the two spin about a common center of mass.

But, isn't this attraction between entities of matter the attraction we have been calling "gravity"?

Then, the rotation of the universe causes particles of matter to attract one another? Then:

THE ROTATION OF THE UNIVERSE IS THE CAUSE OF GRAVITY ?

# GRAVITY!

In the past we have said that bodies of matter are attracted to one another but we didn't know <u>WHY</u>. We weren't even aware that we didn't know why. And now we know why! Our rotating universe gives bodies of matter reason to attract one another, through the requirement of Conservation of Rotational Energy.

Attraction is proportional to the masses of the bodies :: rotational energy is proportional to the masses of the bodies. Attraction is inversely proportional to the square of the distance between the bodies :: rotational energy is directly proportional to the square of the distance between the bodies. And the bodies can be minuscular, or they can be the Earth and the Moon or the Sun and the planets or galaxies and so forth.

Let's define gravity: Gravity is the attractive force acting between all entities of matter and energy by which they form rotating systems to satisfy the rotational energy required of them by their positions in the rotating field of the rotating universe.

To help us further understand the nature of gravity, let us recognize that the only force that can oppose the attractive force of gravity in a steady state of equilibrium is the centrifugal force exerted upon bodies in rotation about one another. An example is a satellite that your brother Brian helped to put into orbit over a fixed position above the earth.

# BEYOND?

If our universe is a rotating, expanding, disintegrating, continually regenerating finite unit, what lies beyond it?

If it is rotating, it must be rotating relative to something else. Is that something else other finite universes?

We have at least one internal hint of what goes on outside our universe. Let us ask: when a galaxy forms at the center of our universe, why does that galaxy move off center to become an "island universe"? Why doesn't it simply stay on center and grow and grow until our universe is all one giant galaxy? A logical answer to this question is that, if our universe is moving in linear or curvilinear motion, gyroscopic motion would cause a newly-formed galaxy to move off center to become an "island universe", or separate galaxy.

If, then, our universe is moving, an extrapolation of our present knowledge leads us to see our universe as part of a still larger order of galaxies of universes.

And where does it all end? We may never know. It may likely be that the more man learns, the more he learns there is to learn, forever and ever.

And where and when did it all begin? It may not have! We find it easier to picture infinity in time as extending on forever forward without end, than we do to picture infinity in time as extending backwards forever without beginning. But we will investigate the beginning of the universe in a later letter.

## O, To Create!

I know that I shall never rate
Among the people who CREATE
The Books, the Music and the Things
That give this simple life its wings.

CHILDREN take up all my time.
I've hardly time to write this rhyme,
With washing, cooking, all the chores
That turn us mothers into bores.

CHILDREN! Bearing them with muffled moan,
Caring for them till they're grown
And, like fledglings they have flown
To bear their own, to rear their own.

And so I know I'll never rate
Among the people who CREATE
The Books, the Music and the Things
That give this simple life its wings,
That give this fleeting life its wings!

Written to Maureen's mother by Maureen's father upon the birth (creation) of Maureen, their sixth child.

# HOW IS THIS UNIVERSE POWERED?

Where would the universe get the energy to drive this recycling centrifuge?

It may well be that there is an antimatter universe, rotating in the opposing direction.

And where would this antimatter universe be located?

It may well be in exactly the same space that our matter universe is, and the two oppose one another, just as two electrons oppose one another, providing the force needed to drive both systems.

So much for the force; where would they get the energy?

If they are exactly equal, and if one is considered positive and the other, rotating in the opposite direction, is considered negative, the total energy of the two is zero.

The universe and the antimatter universe intermeshing within the same sphere? What is the probability of collisions?

Some astronomers believe they see galaxies in our universe intermeshing. They believe there is plenty of open space to avoid collisions.

Let's take a look at the density of matter within our universe. First, let's depict our solar system within a cube one thousand feet on each side--equal to the height of a hundred-story building. If we step into an elevator in the hundred-story building and push the button for the fiftieth floor, we pass forty-nine floors without a planet. On the fiftieth floor, the sun is represented by a bold period in the center of the floor; the nine planets are represented by nine periods, each in its own circle up to one thousand feet in diameter. Add two mouthfuls of smoke distributed in the one hundred storeys to represent the comets, asteroids, meteors and debris. The next star with its planets is ten miles away.

Now consider an antimatter universe, to the same scale, occupying the same space. Its antimatter repels the matter of the matter universe with an anti-gravity repulsion.

The probability of collision between solid bodies of the matter universe and the antimatter universe seems remote. Even within each molecule, the proportions of matter: electrons, protons, neutrons, and empty space are comparable to those in the universe itself.

But what about other than solid bodies? What about mattergy in the form of energy? For example, when I look at a star, two rays of light from the star reach the pupils of my two eyes in two unbroken chains of energy over billions of miles of space. Besides the two rays that reach me, rays radiate from that one star throughout the universe. Each cubic meter of the universe within sight of that star is permeated with light from that star. Even though that cubic meter is completely filled with the light from that star, there is no shortage of space for the light from the next star, and from every star in the universe.

There is no way that an antimatter universe could occupy the same sphere as our universe without occupying space that is totally permeated with the mattergy of our universe.

Just as energy waves are attracted to energy waves, anti-energy waves must repel energy waves so that they can occupy the same space.

Most of us believe that antimatter will search out matter and annihilate it explosively. This belief may be wrong, and may be caused by poor reportage in the newspapers, where most of us learn about antimatter. Explosive release of energy occurs when matter collides with antimatter. In their summaries the reports sometimes fail to state clearly that the collisions require thousands of electron-volts of force to drive matter and antimatter together. This is the force needed to overcome the antigravity of the two kinds of atoms. Just as matter molecules attract matter molecules by gravity, antimatter molecules will repel matter molecules by antigravity.

Let's take a look at the origin of all this, creation.

# CREATION?

Imagination is More Important than Knowledge.
                                    --Albert Einstein

You will be told that the universe started with THE BIG BANG. All the material of the universe was concentrated in one relatively small ball, maybe the diameter of our solar system. Originally, the concentration was proposed as the completion of one cycle of an expanding and collapsing universe. The concentration of energy caused an explosion that sent the matter flying off in all directions and it collected into swirling galaxies. The force of the explosion caused the expansion of the universe, which has since been expanding at a steadily accelerating rate.

I would like to inquire into CREATION, the moment of the first matter or energy coming into existence, not some later time when "all the matter and energy of the universe collected...etc"

And while we are considering the physical universe, I would like to stick to the laws of physics. In the physical universe, for a mass to accelerate requires the application of a force. The equation $F = ma$ {Force equals mass times acceleration} is taught to all physics students as being universally applicable. How is the force of an explosion that occurred eons ago causing the acceleration of galaxies now?

To cause acceleration now, the force must be in effect NOW. In all types of explosions that I am familiar with, a projectile is sent from a gun at muzzle velocity, and loses velocity, does not gain velocity after leaving the gun. In an open explosion (not within a gun), the flying rocks start losing velocity the moment the explosion ends. (Einstein proposed a new force to fill this deficiency, an anti-gravity, repulsive force, but he withdrew it).

(Maureen, EVERYBODY accepts the BIG BANG hypothesis. Everybody except Gold, Hoyle, Bondi and a few other surviving believers in the Steady State theory, who believe that the universe is in a steady state, that new matter is continually being created--in the form of hydrogen--to replace the voids left by expansion of the universe. When they fall back on <u>creation</u> their followers drift away to join "everybody").

I don't mean any disrespect to the Big Bangers. After all, I didn't call this paper "Big Bang? Bunk!" But I'm reminded that when Admiral Perry was presented to the Emperor of Japan and was told to kow-tow, that is, kneel and bang his head loudly three times on the polished wood floor, he refused. I respect the Big Bang theorists; they are, after all, among the leading philosophers of cosmogony: the origin and destiny of the universe. I have learned a lot from them. And I will also respect their hypothesis when they answer the two questions at the top of this letter.

Until then, no kow-tow.

Meanwhile, read on, patiently.

Let's look at our finite but expanding universe. Consider one wave of energy leaving the envelope of the universe, into the void. Refer to the sketch on the following page. The wave causes a wake like that of a speedboat. The wake consists of swirls to the left and swirls to the right. Swirls to the left spin clockwise and swirls to the right spin counterclockwise. Clockwise spinning swirls become the primordia of energy and matter. Counterclockwise swirls become the primordia of anti-energy and antimatter. The algebraic sum of one particle of matter and one particle of antimatter is zero. The accumulation of matter particles will produce a universe and the accumulation of antimatter particles will produce an antimatter universe. The total energy to produce the two universes is zero, except for the initial wave of energy.

Now let's backtrack.

If the one wave of energy is the start of a complete universe, then, if we untangle our previously existing universe, particle by particle, and wave by wave, we should get back to one wave of energy. And when we get back to the original universe, we should get back to the first wave of energy that started it.

Now the question poses: how did that first wave get started?

If we are considering the origin of the original universe, we are asking how did a wave of energy get started in a complete void? By accident? According to the laws of probability, the probability of a wave of energy getting started by accident in an absolute universal void is absolutely zero.

Let's look into the nature of the wave of energy. The spectrum of electromagnetic waves consists of waves of light, waves of sound, and other waves of electromagnetic energy, including thought waves. You may know that thought is accompanied by electric currents from one part of the brain to another, and the magnetism that is a byproduct of these currents can be measured. (You once had an infant brother Timothy, who was brain-damaged by a rough birth, and neurologists measured his brain waves by attaching metal conductors to his bare scalp and measuring the electric current from one part of his brain to another).

In trying to find a probable source of an electromagnetic wave within a complete void, I am positing that the initial wave that started the universe was a thought wave.

Before reading on, note that while considering the physical universe, I insisted on staying within the bounds of physics. At this stage of the discursion, with no physical universe, I do not feel bound by the laws of physics.

A thought wave by whom? It would have to be a thinker from <u>outside</u> the universe because the universe didn't even exist yet. The thought wave created the universe, caused it to exist out of nothing: a thought wave by the CREATOR of the universe.

At this point in the discursion, many books bring in the concept of God, but none so authentically as the Baltimore Catechism. Sixty years ago the Baltimore Catechism taught that God created the universe out of nothing. This present analysis, founded firmly on physics, appears to agree ineluctably with that dogma.

Note: A few years ago (1992) the eminent English astrophysicist, Steven Hawking discerned that the universe is not empty space, but is full of primitive particles of matter and antimatter---. Not too far from our hypothesis. (I don't think he has published his thought in print yet. I received the product of his thought by electromagnetic waves *). *Note: Hawking expressed it on the electromagnetic waves of television while I was lucky enough to have him tuned in.

# Conclusion

When a person has discovered the truth about something and has established it with great effort, then, on viewing his discoveries more carefully, he often realizes that what he has taken such pains to find might have been perceived with the greatest ease. For truth has the property that it is not so deeply concealed as many have thought. . . . Yet it often happens that we do not see what is quite near at hand and clear. And we have a clear example of this right before us. For everything that was demonstrated and explained above so laboriously, is shown to us by Nature so openly and clearly that nothing could be plainer or more obvious.

Galileo,

From "Galileo at work": his Scientific Biography, by Stillman Drake as reported in "Discovery" October 1996.

In an early letter I proposed that the universe is rotating, and I suggested that we inquire into how this fits with what we think about the universe. Let's see how we did.

First, rotation provides us with centrifugal force to power the expansion of the universe. With a stationary universe, Einstein had to invent a new, anti-gravity force, a repulsion, to cause the flying apart of the cosmos. He promptly withdrew the proposal and called it "the biggest mistake of my life".

Applying the centrifugal force to the spacing of the planets, we find that a centrifugal force slightly greater than the gravity attraction toward the sun accounts for the precise spacing of the planets, as tabulated in Table 1, provided the planets originate at the sun at regular intervals. To satisfy the requirement that the planets originate at the sun, we see them as being ejaculated from the sun at equal intervals.

Birth of the planets from the sun lays the genesis for birth of the moons from their planets.

The spacing of the planets gives an insight into the recent discovery of signs of probable life on Mars, namely, that such life probably occurred when Mars was at the same temperature and distance from the sun as Earth is now.

A rotating universe accounts for the elliptical shape of galaxies and of the orbits of planets. Here's an opportunity for a young Kepler to do the math.

The rotating universe puts the expansion and Hubble's Constant right here in our own cosmic backyard among our planets, as detailed in Table 1, and even between our moon and us.

Rotation of the universe offers an explanation of Quasars.

It explains the leap second, the need to add a second every year to correct our clocks for the slowing down of the earth's rotation as it moves away from the sun.

The rotating, regenerating universe answers the question "Where does the energy of the sun and the stars go?"

Is it too high-falutin' to say that the regenerating universe offers a Grand Unified Theory explaining the regeneration of energy to matter and the recycling of the matter into galaxies?

# Conclusion

Unquestionably the greatest outcome of this whole study is the discovery of the cause of the attraction we know as gravity.

# Future

Table 1 offers a measuring rod for locating additional planets.

The existence of the universe and the antimatter universe in the same place is hard to swallow. Any other location appears to me to pose an imponderable plane of shear between them. Maybe if someone reads this book he or she will come up with something better.

A finite universe stipulates a center of gravity, and this means bent light entering our spectrometers, which requires at least some revision of our ideas of Doppler Effect and receding galaxies.

Finding the age of the meteorite from Mars can help us to find the gestation period (Periodicity) of the planets, the time between the birth of Earth and Mars, for example.

April 24, 1977

Dear Maureen,

I am, of course, pleased and flattered to see that you have typed my letters for the purpose of presenting them to a publisher.

Some of my language might seem arrogant unless your readers realize that it was the private language of a father to his daughter and was intended to convey emphasis, not arrogance.

However, I would be pretentious if I did not acknowledge the originators of the ideas presented here. Few, if any, of the ideas are completely new. The order of presentation, the implied interrelationships of the ideas and my inferences are new. And I believe the discernment of the cause of gravity is new.

The order of presentation deserves a table of contents. Since it comes as an after-thought, it should be printed as an after-thought. In this way, it need not interfere with the process of discovery. The discovery is genuine. I did not know, when I started to put some thoughts on paper to you, that I would discover the cause and definition of gravity.

# ACKNOWLEDGMENTS

Herschel, two centuries ago, believed the universe to be rotating.

Newton gave us the laws of gravity, and the laws concerning gravity's antithesis: centrifugal force.

$$2$$

Einstein's law $E = mc$ can be read "energy becomes matter at the speed of light."

Descartes gave us Cartesian vortices and planets that are crusted stars.

Galileo gave us courage to believe and interpret what we see, and, like all of these men, he gave us much more than we can say.

Tycho Brahe gave a lifetime of accurate observations to Kepler, and from these Kepler gave us elliptical orbits.

Jansky gave us a view beyond eyesight, and Ryell found how far that view goes, the extent of our universe.

Lamaitre: the Big Bang theory

Hubble gave us the expanding universe.

Gold, Bondi and Hoyle gave us a self-renewing universe, and Hoyle had the superb audacity to have hydrogen <u>created</u> to do the renewal.

Titus gave us accurate distances of the planets from the sun, and Bode discovered the geometric progression of these intervals.

Alfven believed the Big Bang is a myth.

Afven, Bethe and Gamow represent the countless others who have expanded our knowledge and are continually expanding it towards, but never reaching, Omega.

# APPENDIX No. 1

# BENT LIGHT -- RED SHIFT

Please refer to the color print on the dustcover of this book.

When the ray of light from G is bent by gravity, are the red rays bent to a greater or lesser extent than the violet rays?

The sketch shows two rays of light, each spread out into its spectrum by the difference in effect of gravity on the rays' more energetic and less energetic components from ultraviolet to infrared. (Diffraction).

A narrow slit in the spectroscope picks up all colors of the spectrum of source G.

It picks up violet from the first ray. There is no way it could pick up red from the same ray.

The red it picks up is from the second ray. And each of the colors between red and violet is from a different original ray. Each of the several rays has traveled a different distance from G to E, and each color received in the slit arrives at a different angle. Can this be the cause of Red Shift?

(Note: Although I have omitted references to Relativity in these letters where I felt the omissions would not hurt, I must admit that I have been suspicious that the current explanation of Red Shift, i.e. applying Doppler Effect to the speed of light, is inconsonant with the restricted theory of relativity and so I offer this alternate for your consideration.)

Furthermore, I feel that the idea of our receiving rays of light straight from far distances, and especially from quasars, which I take to be located on the fringes of our pancake-shaped universe, is oversimplified and impossible. If we can't see through our own galaxy, but get only an impression of it as a milky streak across the sky, what chance do we have of seeing straight through our universe, which I am taking to be pancake-shaped like our galaxy and our solar system. Therefore, the only way we can see quasars at the fringes of our galaxy would be in the form of bent light, as shown on the dust jacket.

And so I submit this as my depiction of "Red Shift" for your consideration.

# APPENDIX 2

# KEPLER'S LAWS

Law 1. Each planet travels in an elliptical orbit with the sun as one focus.

Law 2. A line from the sun to a planet sweeps out equal areas in equal times.

Law 3. The ratio of the cube of the radius of the orbit to the square of the period are the same:

$$R_M^3/T_M^2 = R_V^3/T_V^2 = R_E^3/T_E^2$$

# APPENDIX 3

# LEAP SECOND

On the last day of 1977 newspapers reported "The Longest Minute of the Year": the last minute preceding 7 p.m. was assigned 61 seconds in order to keep the clocks of the world in agreement with the rotation of the Earth.

The rotation of the Earth has been found to be slowing down and the correction is required in order that noon may strike as the sun passes through a vertical extension of the meridian.

The cause of the slowing down of the Earth's spin is reported by the newspapers to be a puzzle and a mystery.

If you accept the prediction I gave you: that the planets are receding from the sun, and apply the Principle of Conservation of Rotational Energy, you will have the answer to the mystery. See Table 1, reread a few appropriate letters with this leap second in mind, and ponder.

Some additional experience and refinement of the measurement of time will provide data from which the rate of recession of the Earth from the Sun can be calculated.

By the way, this leap second was no aberration or freak: it has been necessary to add a leap second once every three years since 1968.

# APPENDIX 4

# VULCAN

In the Spacing of the Planets--see table 1 and the equation for the spacing of the planets--the space nearest to the sun is vacant.

Vulcan's place is empty. Where is Vulcan?

Two hundred years ago Herschel, the leading astronomer of the time, discovered a red-hot planet right near the sun. In accordance with the centuries-old tradition of naming planets for gods of Greek mythology, he named his planet Vulcan. Vulcan, the god of red-hot iron and of volcanic eruption, whose orbit, as predicted by the spacing of all the other planets, must be very close to the sun, would be very hard to see. In so tight an orbit, Vulcan would spend very nearly half its time behind the sun and very nearly the other half directly in front of the sun. Its color must be very nearly the color of the sun. Vulcan would be totally invisible except for the brief periods when it would skim just over the sun's horizons.

More than one astronomer has gone blind looking at the sun, with and without telescopes. Hundreds of astronomers over the years have searched for Vulcan, with increasingly better instruments. But Vulcan defies detection. Yet the original discoverer was the leading astronomer of his time, and his reports cannot be dismissed lightly.

And if you will examine the letter on "Spacing of the Planets" and study Table 1, you will find that the place nearest the sun is vacant.

So what has happened to Vulcan? Has it been reabsorbed by the sun? If so, is it spinning in a tight orbit beneath the visible surface of the sun, causing sunspots and magnetic storms? If so, will it be born again? And under what conditions?

Or has the expanding universe started collapsing and returned fiery Vulcan to his fiery womb?

# APPENDIX 5

# LIFE ON MARS

The discovery in 1996 of evidence of earthlike life on Mars billions of years ago indicates that Mars was once at the same temperature that Earth is today, and was therefore at the same distance from the Sun that Earth is today, as predicted in "Spacing of the Planets" given in an earlier letter dated 1978.

# APPENDIX 6

## Corkscrew Path of Light

Scientists have discovered that radio waves travel in a corkscrew path:

Measurements by scientists have suggested for the first time that the universe has an "up" and a "down."

\*\*\*\*

In an analysis of radio waves from 160 distant galaxies, physicists at the University of Rochester and the University of Kansas made the startling discovery that the radiations rotate as they move through space, in a subtle corkscrew pattern unlike anything observed before.

A complete turn of the corkscrew appeared to occur every one billion miles the radio waves travel. . . .

Even more surprisingly, the magnitude of these newly observed rotations appear to depend on the angle at which the radio waves move in relation to a kind of axis of orientation running through space. The more parallel the direction of travel of the wave is with the axis, the greater the rotation. The reason for this remains unknown.

This axis of orientation is not a physical entity, but rather defines a direction of space that somehow determines how light travels through the universe. As observed from Earth, the discoverers said, the axis runs one way toward the constellation Sextans and the other toward the constellation Aquila. Which way is up and which way down, whether toward Sextans or Aquila, would be a matter of arbitrary choice. . . .

New York Times, April 18, 1997 (discussing the research of Dr. Borge Nodland of Rochester and Dr. John Ralston of Kansas).

Try to explain this in the context of a stationary universe.

Explain this phenomenon in the context of a rotating universe.

# APPENDIX 7

# BIRTH OF PLANETS

See Page One of the <u>New</u> <u>York</u> <u>Times</u> of May 29, 1998 for a photograph and article about an observation by the Hubble Telescope of an apparent planet being formed by being expelled from two stars circling each other. Compare this to the discussion at pages 10 through 15 above.

# Glossary

agglomeration: the act of collecting.

antimatter: the opposite of matter.

A.U. (Astronomical Unit): A unit of measure: the distance from the center of the sun to the center of the earth.

asteroids: rocks in orbit about the sun. Most of them travel in a circular group whose orbit is between Mars and Jupiter. See Table 1.

Astronauts: men who travel in shuttle space ships and work in space. Known as cosmonauts in Russia.

binary: Two stars in close proximity, or one star splitting in two, ejaculating a planet.

cataclysmic: extremely violent.

centrifugal: the force caused by rotary motion, in the direction radially away from center of rotation.

centripetal: the force opposing centrifugal force, and holding an object toward center. In the earth-moon system, gravity provides centripetal force.

conservation: saving, maintaining.

constellation: a grouping of stars into a figure such as a cross or animal devised by ancient astronomers for the purpose of identifying them in their places in the sky.

cosmic rays: rays of mattergy from space.

cosmogony:  the study of the origin and destiny of the universe.

cosmology:  study of the cosmos. The study of the universe.

create:  to make out of nothing.

detritus:  broken-up leftovers, usually of stone or stone-like matter.

eclyptic:  the plane in which the planets orbit the sun.

ejaculated:  ejected, thrown out.

ellipse:  an oval-like figure in which the total distance from two foci to each point on the curve is equal. An angular view of a circle. The shape of most constellations.

fault:  a break in the earth's crust. Usually the separation between two tectonic plates which are abrading one another.

focus:  one of two points that determine an ellipse the way the center determines a circle.

galaxy:  a group of a few million stars collected into, usually, an elliptical pinwheel or similar shape.

gravity:  the attraction between each entity of matter or energy and every other such entity in the universe. Word coined in this book as the attraction by which each pair of such entities joins into a couple to satisfy the rotational energy required of them by their position in the rotating universe.

gyroscope:  a relatively massive disc on an axle supported by two bearings, which, when set rotating at high speed, tends to remain fixed in orientation.

magma:  the fluid center of the earth.

mass: the heft of an object. It differs from weight in that weight depends on gravity and so varies with changes in the distance of the object from the center of the earth. Mass is constant, regardless of the distance from the earth.

mattergy: matter and energy. It is matter when it is traveling below the speed of light, and energy when traveling at the speed of light. New word coined in this book.

megaparsec: unit of distance: 3.26 million light-years.

Milky Way: the galaxy in which our solar system is. A milky streak across the sky, our view of the fifty million stars in our galaxy.

minisculum: exceedingly small entity. About as small as an object can be.

nebula (pl. nebulae): from the word nebulous: unclear. A blur about the size of a star to our eye. With improving telescopes in this century, nebulae were found to be galaxies like our own Milky Way Galaxy.

orbit: path, viz: the path of a planet around the sun.

Panagaea: all the land surface of the earth before it broke up into tectonic plates

pendulum: a weight oscillating on a light rod, as on a grandfather's clock.

peripheral: around the edge.

photon: a single wave of light, considered as an object.

radioactive: giving off radiation other than light or heat.

regenerating: renewing.

rotational energy: the energy of an object or a pair of objects rotating about one another, such as the earth and the moon.

quasar: quasi-stellar radiation source.

solar: relating to our sun.

speed of light: $3 \times 10^8$ meters per second.

tectonic plates: the several pieces into which the land surface of the earth is broken.

# SUGGESTED READING

PHYSICS  D. C. HEATH AND COMPANY
 By Physics Science Study Committee.
 Library of Congress Catalog Card No 65-24096.

Herschel, John F.W., Outlines of Astronomy QB43.h57 1902 SH MCLGH

Newton, Isaac, Sir, Philosophia Naturalis Principia - English - also Encyclopedia Britannica.

Descartes, Rene', Discourse on the Method of Rightly Conducting the Reason and Seeking the Truth in the Sciences.

Galileo, Discourse on Two New Sciences 1914 ISBN 0520063600 University of California.

Kepler, Johannes, Publisher: Philosophical Library by Carola Baumpardt with introduction by Albert Einstein LCNN 51993157

Gold, Thomas, Cosmology & Astrophysics ISBN 0801414970 Publisher: Cornell University Press, Owner SH MCLGH QB985.C67 1982

Bondi, Hermann, Sir, Assumption and Myth in Physical Theory, Cambridge U.P. 1967 Owner Blmfld Cl. Jersey City QC71.B63

Bondi, Cosmology, University Press 1961

Bondi, Relativity and Common Sense, Anchor Books 1964

Bondi, Rival Theories of Cosmology, British Broadcasting Corporation QB981.T49 SH MCLGH

Bondi, The Universe at Large, 1960 Anchor Books

Hoyle, Fred, Sir, Astronomy, Doubleday 1962 LCNN62014108/L
    Hoyle, Astronomy and Cosmology, San Francisco W.H. Freeman 1975

Alfven, Cosmology History and Theology, Plenum Press 1977

Alfven, Atom, Man and the Universe, W.H. Freeman 1969

Alfven, Evolution of the Solar System Publisher: NASA
    For sale by Supt. of Documents, U.S. Govt. Printing Office

Alfven, On the Origin of the Solar System, Clarendon Press

Alfven, Structure and Evolutionary History of the Solar System

Hawking, Stephen, On the Rotation of the Universe
              Mon. Not. R. astr. Soc. (1969) 142, 129-141.

# Thank You

I want to thank Maureen's Science teacher, Mrs. Hochstein, for sending Maureen home so full of the questions that led to the letters that became this book. Education is subject to much criticism, and rightly so. I'm put in mind of the police commissioner in the City of New York who saved thousands of lives from murder in his two year administration--imagine: thousands of lives in two years--by one person being personally effective. I feel that people of comparable effectiveness in education could do equally well for the status of education. Mrs. Hochstein could be one of their models.

I can't resist a story about this outstanding teacher. When Brian, one of Maureen's older brothers, was in Mrs. Hochstein's math class he came home one night with the assignment of having his father teach him factoring. My meat. I started to teach him the accepted method of factoring. He interrupted: Mrs. Hochstein had told the class about that method and she warned them that she did not want them taught that method. This led to an argument between Brian and me. I told him that Mrs. Hochstein's husband, Hockey, had been a classmate of mine in engineering for a trimester in wartime Annapolis, and so he was an engineer. Now, an engineer has to know much more math than a junior high school math teacher, and so Hockey was smarter than Mrs. Hochstein. And I am an engineer. Now, if I am as smart as Hockey, and Hockey is smarter than Mrs. Hochstein, then, by simple algebra, I am smarter than Mrs. Hochstein.

Being so smart, I had to know what Mrs. Hochstein wanted. She didn't want Brian shown the concise formulas for factoring; she wanted him to be led by his father up to the open door of the universe and given the opportunity to discover for himself the need for factoring and how it is done. And that's what we did. It must have sunk in because next day Brian was selected to make the presentation to the class. Brian was a painfully honest boy and he didn't stop with factoring, but went on to tell Mrs. Hochstein in front of the whole class all about Hockey and Annapolis and he proved to her--by simple algebra--how his father was smarter than Mrs. Hockstein. Mrs. Hochstein is evidently also painfully honest, because when she got home she told Mr. Hochstein. Before that day, I used to wonder about the meaning and origin of the word brouhaha. Well I don't wonder anymore. Nothing would do but we invite Mrs. and Mr. Hochstein to Sunday dinner and try to smooth the whole thing over. I'm not sure we were successful because we were never invited to Sunday dinner at the Hochsteins.

But thanks anyway, Mrs. Hochstein.

# About Maureen

Yes, Maureen is a real person.

This book was started when she was in junior high school. She has graduated from Edison Junior High School with honors, Westfield Highschool with honors, Dartmouth College with honors, and from The Law School of Chicago University with a prospective husband, Thomas Berg. Maureen and Tom and their son Brendan live in Birmingham, Alabama, where she is a labor lawyer representing working people, Tom teaches law at Cumberland Law School, Samford University, and Brendan, two, is learning to talk and argue cases. I'm reminded that when Maureen was two, I stopped her to give her a job, and she stopped, spread her feet apart and told me "I'm working for mommy. If you want me to do something you'll have to ask mommy first." That's exactly what I had been told to do in the navy if some superior officer told me to do something while I was on a mission for my captain. I told her then "you'll be a labor lawyer someday." She turned on her heel and went about her business for her mother as though that was to be expected.

# About the Author

James T. Kane was educated as a mechanical engineer at the Cooper Union Institute, Rensselaer Polytechnic Institute (R.P.I.), and the U.S. Naval Academy, Annapolis, receiving his degree of B.S.M.E from R.P.I. in 1946 and his Professional Engineer's license from the State of New Jersey in 1950. He has obtained several patents for manufacturing advances and fire prevention. In addition to working in the field of mechanical engineering for more than 50 years, Mr. Kane has studied cosmology and astronomy with a passion for more than 76 years, since he was three years old.

During the 1970's, his daughter Maureen (sixth of his eight children) began asking some questions about the universe, galaxies, stars, and other entities that she was studying in Earth Science. Her curiosity prompted Mr. Kane to write in letters some of the answers that he had reached in his own personal studies. Writing these letters eventually led Mr. Kane to some startling discoveries, which are laid out in this book in a manner to be readable and entertaining for everyone.

www.ingramcontent.com/pod-product-compliance
Lightning Source LLC
Chambersburg PA
CBHW030855180526
45163CB00004B/1586